Go Design It!: An Informative Guide for Effective Design Engineering

Expert Advice from an Engineer Par Excellence

Table of Contents

Chapter 1-Design Objectives

Design engineers are the artists of the engineering world. We bring revolutionary ideas to reality using our technological innovation and expertise.

The success of a design engineer rests on several factors. One reason why most designers are successful in creating great designs is that they are willing to ask themselves the right questions.

Asking the right questions has always worked for me because it allows me to concentrate on a specific aspect of a problem. Design engineers work well when they focus their brainpower on solving problems with the application of science and mathematics.

Most of the questions should circle around the design objectives. Clarity regarding the design objective helps in de-cluttering even the messiest of ideas.

Design objectives are the functional and non-functional qualities that glue the design together. These usually revolve around the features the design wishes to achieve. The objectives also serve as a great tool to measure the progress of the design.

The understanding of these objectives is essential. A clear image of the output enables you to choose the right input.

The right combination of information, materials, and energy gives you the output that you are looking for.

The functions that are responsible for turning an input into output are often very complicated. It can be challenging to simplify these functions into an easy to understand the procedure.

It is the job of the design engineer to engage in a process of modeling that makes complicated problems easily manageable.

The first question that you need to ask yourself is what you want your design to look like, and what purpose does it serve? This question revolves around the basics of form and function, and how they come together in a deliberately designed manner. The third most important question is, who is it for? The end-user, or the client, is the person you must always strive to appease.

Design Goals

The questions you ask yourself should follow each other periodically. It is easy for beginners to confuse their design objectives and struggle throughout the whole process.

After you are done setting your design objectives and have a relatively clear picture of what you want to achieve, you can then move on to the fun part.

As a designer, I find this the most challenging and yet the most entertaining aspect of the whole design process.

However, when choosing a design or style for your product you need to clarify what it is you are trying to achieve. In certain creative environments, aesthetically pleasing designs will be all you need to worry about. However, it is more than likely that your design will need to be a combination of aesthetics and functionality. Since you use mathematical and scientific principles, which are often interlinked, it is always a good idea to start by clearly defining the geometric properties that you are going to follow.

This is another question that can lead to endless possibilities. The endless possibilities will lead to the formulation of different ideas and different solutions to the problem. The plans and possibilities allow you to create designs that are based on these solutions. With all the designs that you have in your mind, you can easily choose the one that suits your idea the best.

The design leads to the creation of prototypes that you can use to test the effectiveness of the product. The final goal of the design process is to ensure that all elements combine to produce an effective product.

Operational Objectives

Engineers use specialized expertise to provide technological operations. As an engineer, you should be aware of what you want to achieve from your objectives. The achievement of your objectives depends on how well you plan for them.

Going by the explanation presented above, it is easy to say that the operational objectives you set for yourself are reflective of how well you understand the project.

You see, unlike goals, objectives are specific milestones that must be met. The parameters for objectives you set for the project revolve primarily around "who, what, when, where, and how."

Rather than thinking about design in its entirety, it is always easier to focus on small bits and pieces that together make up the whole project. Just like building a puzzle, these small tasks done little by little will result in the completion of the project.

The success of your project design depends on how early you recognize the project goals. This is why it is important to have all the necessary information before starting.

Since identifying and categorizing your objectives is integral to the planning phase, I have made a list of factors that you should consider when setting these objectives.

1. Functional/Operational

As I mentioned before, a clear and precise understanding of the functional and physical requirements of a project is a must. Once you create a rough design or plan an end product, the idea must fulfill your cognitive needs. A lot of time and money can be saved by correctly achieving the requirements of the design as early on in the process as possible.

2. Cost-Effectiveness

Like all other plans, your design plan should be cost-effective too. You should ensure that the production cost of the design and the creation of the final product requires a minimal cost. This can be done by effectively designing for specific manufacturing processes and material selections.

3. Sustainability

The last thing that you need to ask yourself is regarding the sustainability of the product. Sustainability is a must because it ensures the longevity of the final design. This is where you ensure your design will withstand the test of time in the environment that it will be used in.

4. Aesthetics

The next thing that should be a part of your design objectives is the aesthetics of the product. As mentioned earlier, it may not be necessary to make every product aesthetically pleasing. That is something that you will need to decide based on the circumstances of your design.

5. Accessibility

The first thing that you should consider is the accessibility of the design you are creating. You need to figure out the availability of the materials required for the development of a product.

Setting the design objectives helps give you a general idea of the manufacturing method you will adopt for designing the final product.

The creation of your design relies on the four W's of design. Once a prototype has been made it will allow you to test and check some of the objectives such as aesthetics and functionality. Some basic questions you should ask yourself before beginning are as follows.

What are you trying to accomplish?

What do you want it to look like?

What function do you want it to have?

What materials are required to make it?

I know that I have already discussed most of those questions before. However, to achieve a fool-proof plan, you need to devise a systematic strategy according to the questions mentioned above.

What Are You Trying to Accomplish?

The answer to this question depends on the functions you have in mind when designing a final product. As a design engineer myself, I get a list of requirements or even a basic napkin sketch, which is often unclear. However, as a professional, it is my responsibility to transform these rough ideas into a professional engineering design.

The first thing that you need to ask yourself is what are you designing and for whom you are designing the product. This is an introspective question that can lead to multiple answers.

You need to question the dynamics of the product. The requirements of a completely new design can be different from that of one that is just an addition or revision to an existing design. The right mix of goals and functions is enough to help you understand the requirements of the user in both formal and technical terms.

Geometric constraints and ease of manufacturing are part of the goals and functions the designer wishes to achieve. These are strict limits that a design must meet to be acceptable. They also act as guides for how the designer can go about producing the parts for the product. You can create a list of attributes that can be used to identify the objectives of the design.

Once you have clarity on the objectives of the design, you can move on to create a hierarchical objective tree that provides a mechanism to all interested parties. Asking the question of purpose makes it easier for a designer to visualize the design in mind. This is a great way to start things off and set up a procedural design framework.

What Do You Want It to Look Like?

Now, this part of the analysis and questioning requires the use of imagination, graphical tools, and design software.
One tool that may help you is the Morphological Chart (see Figure 1.1). You can use it to visualize the size of the design space. The entries in the chart can help you in describing the means through which you can achieve a particular design.

On the left side of the chart, you can list down the functions, while on the right side; you can jot down the different mechanisms that you can use to achieve these functions.

Additionally, you will also get a list of specifications and metrics that will further help in explaining what the product should look like. See the example below.

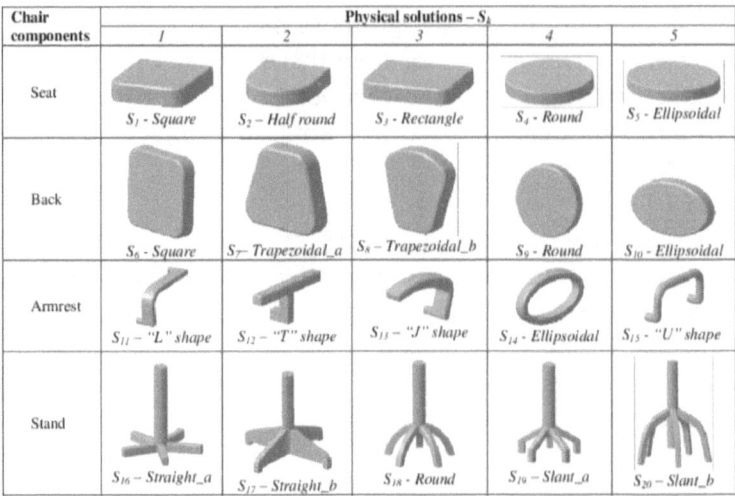

Figure 1.1

You can also use good old-fashioned imagination, creativity, hand sketching, and experience to determine what you want it to look like.

What Function Do You Want it to Have?

Once you have figured out the goals and desired look of the product, you need to move on to the next question in the study.

12

Rather than getting overwhelmed by trying to imagine a fully completed design in advance, you should begin by focusing on one small section at a time. This will allow you to approach the problem little by little and methodically find solutions until you have achieved every objective of the design.

Moreover, the functions and the overall features of the product are usually specified in very clear terms. These specifications and design metrics are often enough to understand the end goal and the functions the producer wants the design to feature.

One way to systematically determine the desired function of the design is to use a function-means tree (see Figure 1.2). The function means a tree is used as an illustration that elaborates on the primary and secondary design functions. You can represent the functions as rectangles and the means of achieving the functions as trapezoids. The design of the means tree is hierarchical, which means that some functions will also require secondary functions. Below is an example template of a function that means tree. The function is what you want it to do and means are the solutions or ways to achieve it.

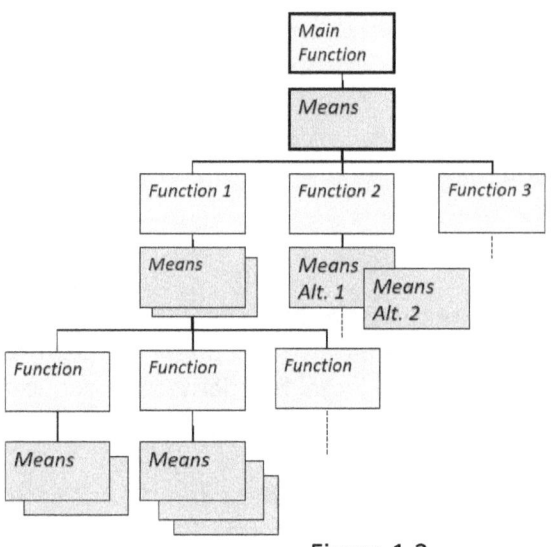

Figure 1.2

What Materials Are Required to Make It?

The material you select largely depends on the function of the design and the process in which you want it to be manufactured. The 3 primary groups of materials are metals, ceramics, and polymers. The science of materials is beyond the scope of this book but there are some basic properties for each of these 3 categories that you may consider for your design intent.

Typically, metals are stiff, ductile, and good conductors of electricity. Metals can be machined or fabricated to desired specifications and are so widely used that many shops around the world have the expertise to make metal parts of all shapes and sizes. Metals can be very corrosive resistant and strong. This is an advantage for outdoor applications or places where external chemicals may be a factor.

Ceramics are stiff, strong, and bad conductors of electricity. They can also be very brittle and susceptible to fracture. Some uses for Ceramics are cookware, cutlery, and automobile parts.

In the polymer's family are the plastic and rubber materials. Plastics are typically lightweight and flexible, and in certain cases can be very strong. They are the material of choice for injection molding parts for many different industries. Also, plastics are widely used and there are many shops that can form them into all different shapes and sizes.

It is up to the designer to determine what qualities are needed for their application and to do the proper research to decide what material best fits their design needs.

Chapter 2- The Foundation of Design

Before going to the next stage of the design process, it is imperative to have absolute confidence regarding the project.

A favorable analogy that simplifies the process of designing is a puzzle. The process of engineering designing is similar to that of a puzzle.

Like in a puzzle, the designing process should make use of pieces that you are aware of as you work towards the pieces that you are not aware of. Using the little bits information you do know to complete the designing process is a logical way to approach it.

When planning the design of a product, you should contemplate the shape and size of the product you are about to design. Additionally, great attention needs to be given to the method in which it will be manufactured. This will undoubtedly drive some of the features of the parts you are inventing. When in doubt about the manufacturing-driven features, it is best to consult professionals of that trade and seek their advice. This way you will ensure that what you are designing does not become unnecessarily expensive to make.

There does not need to be absolute clarity in your head regarding the shape and size of the design you need. On many occasions, this will take a trial and error approach but the more clearly you understand your design objectives in advance the quicker you will realize the goals you are trying to achieve. There is no harm in a design not working out according to the expectations; not trusting and backing your instinct, however, is harmful.

The questions should revolve around the shape and size of the product, materials used, and the aesthetics that you want to follow. Given the various options available, you can decide what shape and size you want the product to have.

The material that you will use to manufacture the product will have a big impact on the way you design the parts. Certain materials can only be efficiently manufactured in certain ways. You must consider this when creating the geometry of your parts.

To begin the designing process and to test the stature of your design, it is imperative to analyze the dimensions. The analysis will help in eliminating the variables of the design.

You will undoubtedly have certain dimensions or sizes that you must workaround. This can be an excellent starting point.

Dimensional Analysis (Introduction)

Engineers and physicists use dimensional analysis to reduce the complexity involved with the fundamental equations. This allows the engineer to describe the behavior in the most economical manner.

DA provides the users with the benefit of savings during the stage of experiments and to help in scaling the standard experiments.

In the world of engineering and science, dimensional analysis is the process of analyzing the relationships between the different physical quantities. The analysis revolves around the identification of the base quantities and units of measurement. Once you have an idea of the base quantities, you can track the dimensions as calculations while performing comparisons of the base quantities.

The main advantage of using dimensional analysis is that the process reduces the number of variables in a problem. The process involves combining dimensional variables to combine non-dimensional parameters for the user.

The use of dimensional analysis is significant in the process of manually designing a model. The process plays a critical role in understanding the model and discovering the variables and constants on the show. By exploring dimensional analysis and understanding the process involved you set the tone for the upcoming methods.

The main question that concerns design engineers in the initial process is the limited information that they have to work with. The limited information should not hamper the design process, as the engineers should follow these steps.

Design Around Constants or dimensions and geometry that are set in stone.

Some engineers might argue that dimensional analysis is the first step of the process. As mentioned before the role of a design engineer is quite similar to that of a child solving puzzles. Like the child, design engineers must work around the puzzle to finding the best design. Given the nature of the software tools available today the design engineers have the liberty of going back to the drawing board in case the design does not satisfy the requirements.

You see, the role of a design engineer is to use engineering sciences to convert resources optimally for them to meet an objective. I have already discussed the processes that a designer should follow before the designing process starts. Your planning before the designing process starts is tested during the process. Improper planning will affect the key fundamentals of the designing process.

Synthesis, analysis, and criteria are among the fundamental elements of a design process. As mentioned before without these fundamentals it is challenging to set objectives for a business.

Determining Constants

Unlike variables or things that you must invent to make your design work, constants are fixed dimensions or constraints. This here is the first part of setting an objective for a design. Now that you have an idea of the design requirements of your job it is time to determine the constants that gives a direction to the process of design engineering. The process of determining the constants and variables should immediately follow the idea of setting guidelines for the design process and understanding the objectives of the design.

All products have constants that allow the engineer to work towards the design objective. This is where the puzzle theory comes into effect, the constants work as the pieces of the puzzle that know so you can find the other pieces that you don't.

Finding the constants can be both challenging and fun depending on the nature of the product that you are working on. Some products have constants that are visible quite clearly while others require some mathematical calculation. The calculation helps the user to determine the properties of the constant in light of the dimensions the product holds. Once you have identified some constants you can start designing around them using your imagination and the requirements you wish to meet.

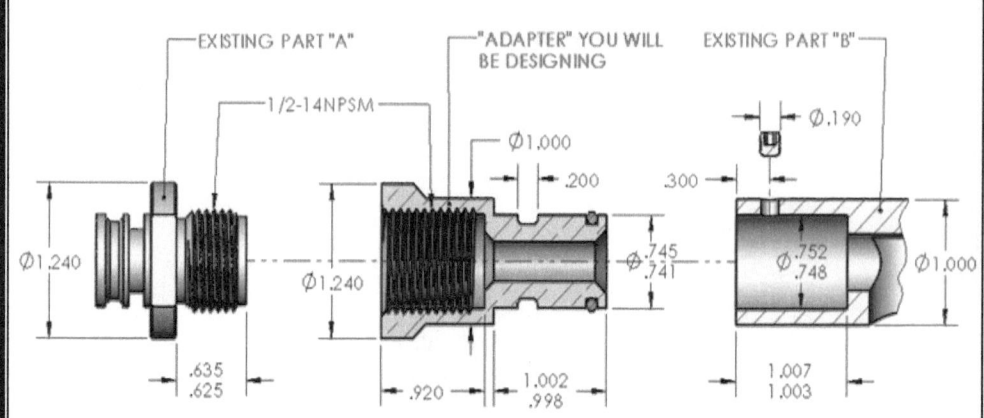

Figure 1.3

For example, let's say you have 2 existing parts, part A and part B, and you need to design an adapter so they can be fitted together (see Figure 1.3). Your constants from part "B" would be the ⌀.752/748,the1.007/1.003, the .300 position of the set screw, the ⌀.190 of the set screw, and the ⌀1.000. Your constants from part "A" will be the ⌀1.240, the .635/.625, and the ½-14NPSM MALE THREAD. As you can see from the image, the dimensions of the constants drive the dimensions of the new adapter. The adapter will slide into part B and be secured by the set screw. The adapter and part B will have matching outer diameters. They have been dimensioned in a way that interference is not possible. The same goes for part A as it will screw into the adapter. They will also have matching outer diameters. In this case, the length of the threaded portion on part A is much shorter than the internal thread on the adapter. That was a subjective decision made by the designer for one reason or another.

Finding the factors that are constant allows the user the liberty of adding on things that are unconstrained by dimensions and other limiting factors. Once you figure out the presence of a constant you can work the design and incorporate all the other factors to the design like a puzzle. By working on the design like a puzzle you can reveal the essential factors that will help you in achieving the final target and design of the product.

In the early stage of the design process, the constraints tend to be challenging. This is especially true when working on large assemblies with many components. However, as more and more constraints are added to design the variables begin to narrow until you have completed the puzzle. Well, what if you are starting a design from scratch with no existing parts or constraints? The next section will discuss how you can handle this.

Shape, Size, and Function

One challenge that all design engineers face is the identification of constraints and constants. If the design in hand does not have any current mating parts you should choose constraints based on the shape, size, and function that you have determined it requires.

Usually, designers are provided with instructions from a manufacturer or inventor including the description of the product look. The shape that they need helps the designer in finding the best possible solution. However, in cases where there is an absence of predetermined designs and shapes, you can use your own imagination and knowledge and consider how you want the shape to look.

It is not necessary for every product to be aesthetically pleasing and attractive to the eye. Some products don't require such aesthetics; a good design engineer can differentiate between the design requirements of different products.

Once you have found the basics regarding the shape, size, and function of the product you can use these as your constants to design around. The whole idea of the process is to find the missing components and then invent them.

Designing, a Game of Patience

Often designers get overwhelmed when dealing with designs and the different components. This can end up stifling their creativity and thinking and cost valuable time.

The main reason for the confusion is the daunting task they have at hand. Some may make the mistake of attempting to picture the end goal before they even start the process of designing the product. You would never try to build a house all at once. You would build it little by little until you have a finished house. This is how you should approach design.

As a modern engineer, the use of a software tool is mandatory. One beauty of CAD and the real advantage that it brings to the user is that you can update the current design and revise as many times as necessary to get the results that you are looking for.

The role of the CAD is to aid the creation and modification or optimization of a design. The software naturally comes with all the necessary tools required. The software is designed particularly to increase the productivity of the user. The documentation of the design helps the designer with communication and enables the creation of a database for the manufacturing process.

SolidWorks is a computer program that assists in the development of a model. This computer-aided engineering program is particularly popular among users. One major advantage of CAD is it allows for countless iterations. It also provides a geometrically and visually accurate model for you to analyze.

Choosing the Correct Design Environment in Your Design Software

Before drawing any components, determine what sort of design template you need. The template that you decide to work on depends primarily on the particular type of product you are trying to design. The template should be set up with the appropriate units, drafting standards, required measurement precision, and any other personal preferences.

As a designer, it is easy to shape the design according to your understanding and desire. However, you must always consider in what part of the world your design will be manufactured. The location where the design is produced is significant. Different countries will have different standards and units which you must identify and adhere to if you want a successful translation of your design intent. If it will be made in a country that uses the metric system then it may be wise to design using metric units. If the part will be made in the U.S. then inches will be units to use. Also, consider the precision at which it will need to be made and make sure your decimal places match the manufacturer's capability.

There are times when dimensioning infractions is advantageous. Times, when you may consider using fractions, is when designing something that will be built in wood or through a process where measurements will be taken using a tape measure. This way the builder can easily correlate and then apply the measurements from the design while building it.

A software tool such as SolidWorks will assist you in finding the right fractions for the mechanism. The use of fractions also depends on the extent of tolerance available. The tolerance is the permissible limit of variation. Most tape measures have increments of 1/16" or 1/32" and thus that is their measuring precision. I advise you to only use fractions when your dimension scan varies by this amount.

No design engineer wants the manufacturer to read the engineering drawings and models and interpret them incorrectly. These complications can be drastically narrowed if you give thought and consideration to the things mentioned above.

Dual Dimensioning

One way to help avoid confusion and make your designs universal is to use dual dimensioning on your blueprints.

Dual dimensioning is the process by which you display a design using two sets of units, such as inches and millimeters. This will make your design interchangeable for any country. The presentation of the dual dimension depends on the design of the product and on the information that you are trying to present.

Currently, design engineering experts either place dual dimensions side by side or give them a stacked representation. This allows them to present the typical information in both inches and millimeters.

When using dual dimensions on a blueprint it helps to clarify by having a note on the blueprint that states that one set of dimensions is in a different type of unit (see Figure 1.4). Typically, one dimension is placed in Brackets and the other is not. It is important that the tolerances you call out for both dimensions must be equivalent. This way you will get parts that function properly regardless of which unit of measurement is used. Notice that in the tolerance block, I have specified a tolerance ±.005" for .XXX or ±.127mm for .XX, which are equivalent.

There are different drafting standards that aid the process of designing the product. While software such as SolidWorks and CAD do support the use of dual dimensions there is no standard that accepts dual dimensions. Standards such as the

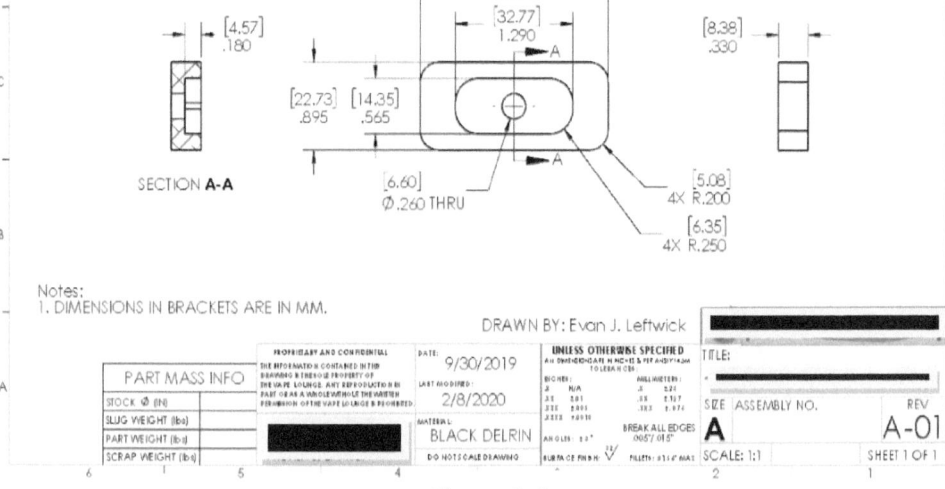

Figure 1.4

ASME Y14.5M - 1994 do not provide the user with any rule regarding dual dimensioning. These different standards help you during the process of drafting and give specific guidelines on how the dimensions are to be interpreted on a blueprint. The guidelines for interpretation make it so there is no confusion as to what is being asked of the manufacturer.

Chapter 3- The Design Process

Now that you have set the foundation for the design and a template that you will follow, it is time to start the design process.

Remember that the basics of the design process are pre-determined. However, you also have the creative liberty of exploring the different options available. There will be times when you will have to scrap the whole project and start all over again. This is normal in the design process as designers take several attempts to get the best results and deliver the product that is expected of them.

When you start the design process, you begin by working on a limitless 3D drawing area, which is called the model space. you should start sketching around the origin in the model space. You need to decide the unit of measurement and ensure that the drawing matches the unit set throughout.

You can choose from the different units available. Once you have the most convenient unit, begin drawing at a scale of 1:1.

Using the origin at the center of your parts helps you utilize the symmetry of the geometry (see Figure 1.5). The symmetric nature of the design will allow you to use common tools such as mirroring patterning, revolving, and cross-sectioning. These things will save you valuable time.

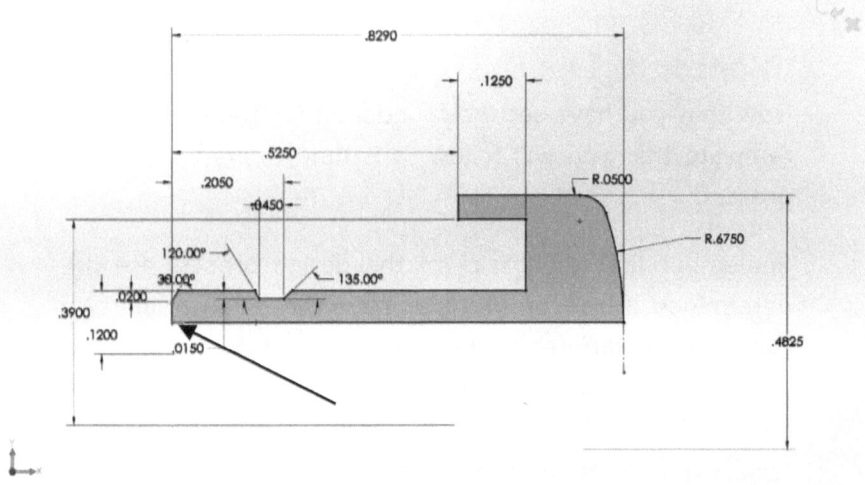

Figure 1.5

The mirror tool will provide you with the power of drawing only one half of a part and then easily making a mirror image copy. You will then have the full part. All you have to do is define the axis of mirroring. The tool will create the second half of the part which can save a lot of time.

The pattern tool will allow you to sketch one portion of a large repeating pattern and then generate the whole thing. It can save tons of time because` you just draw one, then use

28

the tool to repeat it as many times as you want at whatever location you want it to be.

In most CAD tools you will first create the 3D model and then create the dimensioned blueprint after. The best time to determine special fits and tolerances is while creating the 3D model. This will save time later on. While creating the 3D model, make sure it will fit properly with its mating parts. Use annotations for specific dimensions that fall outside of the standard tolerances in the design space (see Figure 1.6). This will aid you later and speed things up when making the blueprints. Blueprints can be tedious and time-consuming so anyway that you can make the process easier will help you and your sanity. By doing it this way, you will have all the heavy lifting done when you are finished with the 3d design. The same strategy is beneficial when designing threaded parts. Making notes and determine thread callouts during the design process. When it comes time to make blueprints you can just effortlessly plug and play your predetermined tolerances.

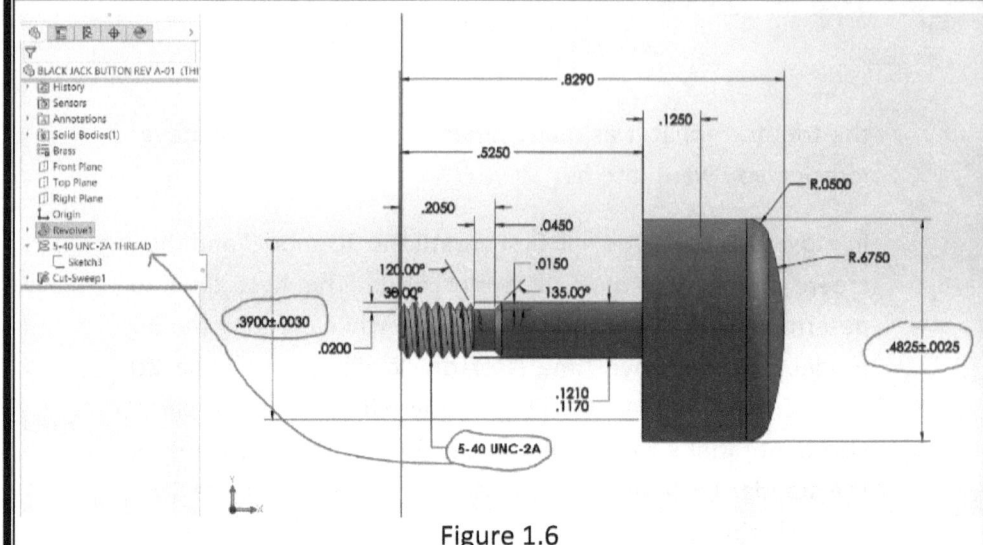

Figure 1.6

During the design process, you might come across parts with a lot of design features. In most CAD software each feature is labeled in some sort of list or design tree (See Figure 1.7). These lists of features can get confusing if you do not make it a priority to keep it organized and legible. It is important to clearly label the features of design with names that make sense to you in a manner you will easily remember. This way, even if you come back to a design years later and need to make edits you can easily find and know what each feature of a part is.

I know getting through the pre-design process can be tedious, but the more you pay attention to the details, the easier it is for you to design the other aspects of the component.

30

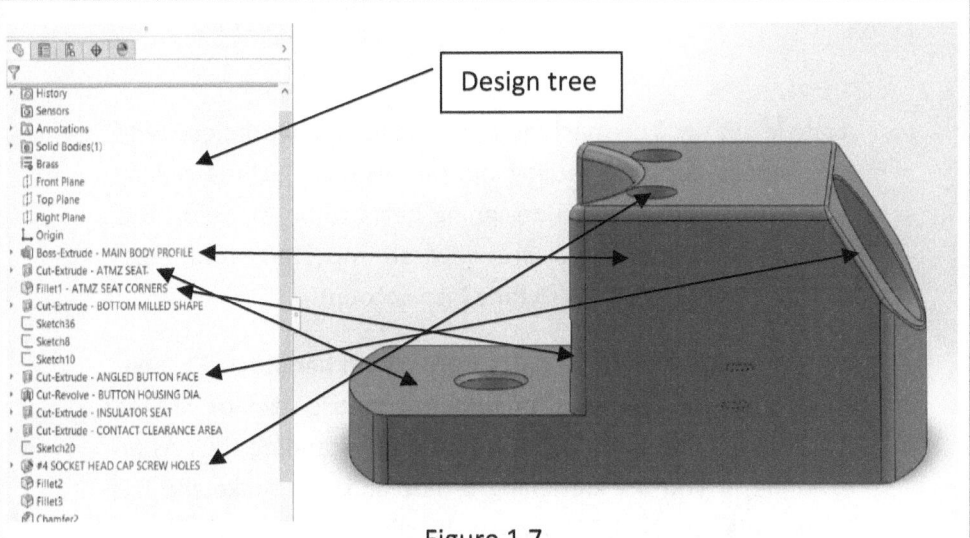

Figure 1.7

Another aspect of creating the geometry of the parts is considering the process in which they will be manufactured. Many times, this will drive the geometry of certain features. It is time to move onto the design considerations. These are the considerations that will allow you to focus on different questions regarding the product.

There are many different forms of manufacturing. In this next section, I will discuss some of the most common ways in which things are manufactured, and I will highlight some important things that you need to consider in order for your components to be made with ease and precision.

1. CNC Mill Turn Lathe

Mill Turn Lathe is one of the most popular types of CNC machines out there. This machine is adept at multi-tasking and can provide the user with complex workpieces with just a single set up. The machine can transform a workpiece with the help of rotating tooling operations that include milling and cross-drilling. When designing parts to be made on this type of machine there are several important and cost-sensitive considerations to take into account.

Mill-turn lathes machine parts from round bar stock. This bar stock comes in standard fractional or metric sizes of varying increments. Costs can be saved by understanding this. As an example, if you are designing a part and you make the OD (outer diameter) 1.26" which is slightly larger than the fractional size of 1-1/4" you may be wasting money. If you are able to reduce the size of a 1.26" part to 1.25" to save material, then it would be wise to do so. In this example, a 1.26" part would have to be made from 1-3/8" bar. Just to cut the bar down to 1.26" from 1-3/8" you are already scrapping about 8% of the material. If you are able to reduce it to 1.25" then you can save those unnecessary material costs.

Since Mill-turn machines are able to do milling, turning, and drilling in different axis they are very versatile. It is important to consider in what way the part will be held or fixtured so that it can be made. When possible, leave a cylindrical surface uninterrupted by milled flats that would make it difficult to fixture on a sub-spindle or for secondary operations. This will ensure the Machinist has an easy and solid surface to grip on to do any further work required. This also goes for thin-walled parts. Try and avoid very thin-walled parts, this will only cause machining issues and drive costs up.

In our modern manufacturing world, it is not uncommon to have complex 3d surfaces machined on to parts. With CAD/CAM this has become much more possible but is still time-consuming. Consider the costs and extra time it may take to machine advanced 5 axis surfaces. If it can be avoided, you should do so in the name of cost savings and ease of manufacturing.

On cylindrical surfaces that will be turned be aware of fillet radii (See Figure 1.8). Most metal turning tools have at least R.004" on their cutting tip which makes the tool stronger. It also makes achieving a smoother finish easier. Be conscious that asking for sharp internal edges can make machining more difficult. If sharp corners are necessary, make sure you specify that. Account for this fillet radius on parts that may be mated together where it may cause interference in the fit or function.

Figure 1.8

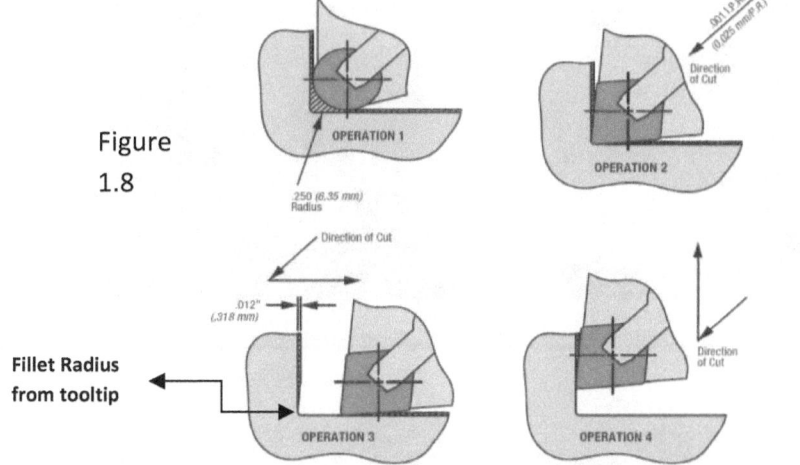

Fillet Radius from tooltip

33

Also, avoid the development of unnecessary sharp id features that would require broaching. Make sure that the corners have a large enough radius so that they can be milled with a milling cutter of reasonable size (see Figure 1.9). For example, say you have a square internal feature that is 1.5" deep and the corners only have a radius of 1/16". Well, then you are asking them to use a 1/8" endmill to cut 1.5" deep. That would be a very small endmill for a cut that deep and could slow down the machining while driving costs up. If square corners and broaching are necessary, then leave some relief or open space behind the feature. That way there is an empty space for the material to escape to when it is being broached out. That relieves pressure on the broaching tool and helps avoid tool breakage.

Figure 1.9

CNC Milling

The second machining option that you can consider is milling. Milling lends itself to parts whose outer geometry is mostly square. Many of the things mentioned in the mill-turn lathe section apply here as well. When using the option, make sure that there are no sharp internal corners that are impossible to make with a mill cutter (see Figure 1.9).

You need to ensure that you leave room for large endmills in areas that a lot of material will be removed. Having a large corner radius and avoiding deep narrow pockets will help this. This way they can use a large rigid tool to mill away a lot of material. This will reduce part cutting time, and costs. Many make the mistake of designing deep small pockets. These are difficult to cut because it would require endmills that are long yet small in diameter. They will be non-rigid and tend to chatter, wear fast, and break easily.

To get the best results and prices out of all types of CNC machining you should be realistic about the tolerances you are asking for. Understand what is possible with these types of machines and try and stay within those limits. You can save a lot of money by making these considerations.

Molding of parts

Another popular way of manufacturing parts is by molding or casting. This is the process of forming a malleable or liquid material into a cavity of the desired shape. It then hardens in that desired shape and can be removed and used for its intention. There are multiple factors that you need to consider when you are designing for the molding process.

One of these factors is designing the part in a way that it has uniform wall thickness throughout. When pouring a liquid or molten material into a mold, having uniform wall thickness ensures that it will cool or harden uniformly. This helps eliminate part shrinkage or warpage after it is removed from the mold. This will ensure the part is much more accurate, and easy to produce. There are times where it is hard to achieve uniform wall thickness due to strength issues. You may want a certain area of your part to be thicker so that it is stronger. In this case, you can consider using ribs of the same uniform thickness (see Figure 1.10). Ribs are cross members that run horizontal or vertical across the area for strength. You can achieve strength and uniformity by doing this.

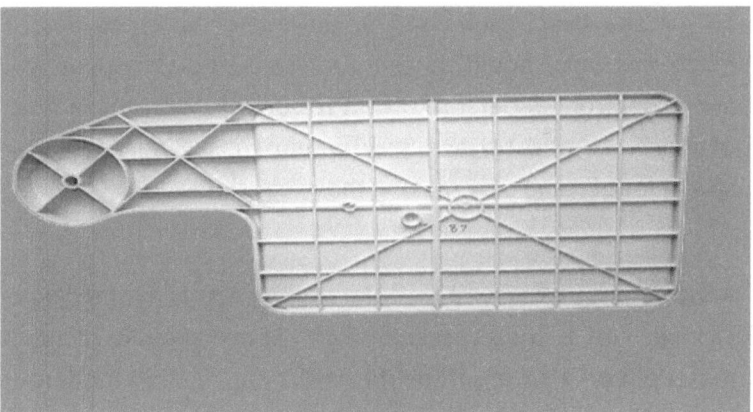

Figure 1.10

Another important consideration is having sufficient draft angle. A Draft angle is a taper, measured in degrees, on the surfaces of the parts (see Figure 1.11). A sufficient draft angle helps in reducing the friction between the cooling and the finished part with a side of the mold. This allows parts to be extracted from the mold much easier. It is a very important and necessary factor in molded parts. Using the correct draft angle reduces the likelihood of damage and wear and tear of the mold. The process also helps in the creation of a smooth and uniform surface finish.

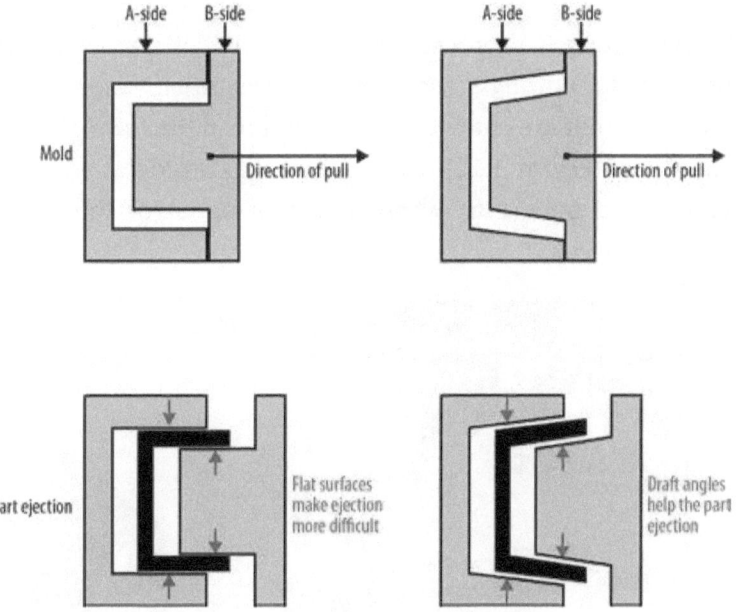

Figure 1.11

Having radii on the corners of your parts is another important aid in the molding process. You need to ensure that the corner radius is as large as possible. This will increase the flow of the material and improve the distribution of its finished strength.

A lot can be learned by simply communicating with the company doing the molding as they are the experts. Most of the time they will gladly offer advice that will make yours and their jobs easier.

Water Jetting Considerations

Water jetting is a process that uses a jet of water at high pressure for cutting a wide variety of materials. Usually, the materials that are cut with a water jet are in the form of a flat sheet (see Figure 1.12). Here are some considerations that you should adhere to when designing parts for the water jetting process.

Figure 1.12

The width of the water jet stream varies from machine to machine. Thought should be given to how that width will affect small details of your design. For example, since the shape of the stream is that of a cylinder, it is impossible for it to cut perfectly sharp corners. Also, let's imagine that the width of the jet is a diameter of .05". If you are trying to design a small feature that has a cutout of only .04" wide it will overcut. There is no way for the jet to cut in between a .04" wide area because it is smaller than the jet itself.

Another thing to think about is how the jet will spread out as it travels through the sheet of material. The thicker the material that it must cut through, the more distance the jet has time to fan out. The dimensions on the top of the part might be right on the money but the bottom side will likely be slightly undersized. This is due to the taper left by the spreading out of the water jet. Usually, this is a very small amount. It isn't advisable to water jet components that require a high degree of accuracy.

Since water jetted parts are usually cut from large flat sheets it is important to think about how they will be arranged on that sheet to maximize the amount of usable material. Designing your parts in a way that they can be tightly and efficiently arranged can help save money. This will give you more area per sheet that is usable space.

Sheet Metal Considerations
Sheet metal is widely used in modern-day manufacturing. It can be a very affordable and versatile process for many

different types of products. Therefore, it is important to have a good understanding of how parts should be designed for manufacturing using sheet metal.

Sheet metal parts are just that, they are sheets of metal that are drilled, punched, stamped and bent to the desired shape. The flat sheets that they are made from come in standard thickness sizes. These thicknesses are designated by gauge numbers. For example, a copper sheet that is .043" thick is known as 19 Gauge copper sheet. Past a certain thickness, the material is then called a plate. When designing sheet metal parts, be sure you are specifying a standard gauge size. Information on these thicknesses can easily be found on metal supplier websites or the internet. The thickness that you decide on will be unique to your specific product based on what you need it to do.

When sheet metal is bent from its flat form it actually stretches and compresses the metal at the bend points. Because of this, to achieve the proper dimensions of the bent part, the flat piece will need to be smaller than the folded one (see Figure 1.13). The size of the flat piece needed to create a bent part can be calculated from the dimensions of the bent part by using the bend allowance, bend deduction and the K factor. The formula for k factor will help you determine the elongation of your metal sheet. Having precise information about the K factor will help you in calculating the flat patterns stretch amount during the process of bending.

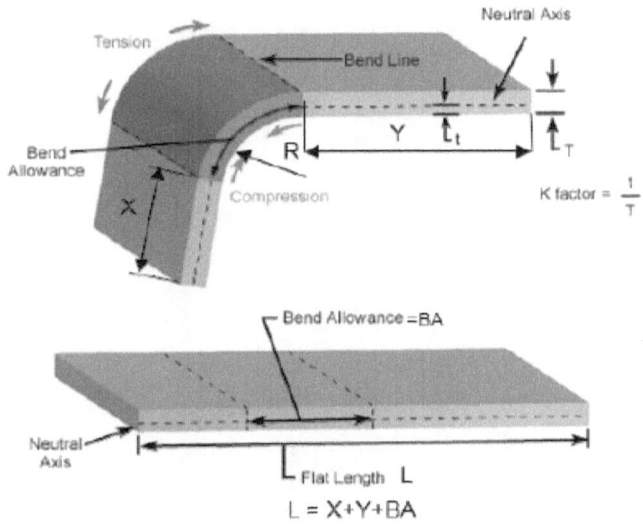

Figure 1.13

Sheet metal bending is done by sandwiching a flat sheet between two pieces of metal that are the desired shape (see Figure 1.14). One piece called a punch applies pressure and forces the sheet into the die. When bending a sheet metal part there will always be a radius on the side opposite the punch. Even if the tip of the punch is perfectly sharp you will still get a radius equal to the thickness of the sheet. Understand that you will never be able to get a perfectly sharp corner on both sides of a bent sheet metal part. Also, when bending certain features such as flanges. It is good practice to cut away a small section on each side of the flange so the metal doesn't crack and tear.

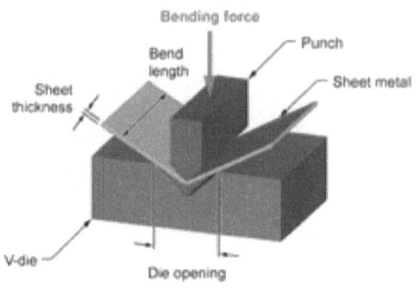

Figure 1.14

Software programs such as SolidWorks have specific tools for Sheet metal design that considerably simplify all the things discussed in this section.

Considerations for Designing Threaded Parts

Screwing components together is the most common way of fastening parts to one another. That is why it is imperative to understand how to properly design threaded parts. Male threaded parts such as bolts, screw into female threaded parts such as nuts. Male threads are threads on the outer diameter, whereas female threads are on the internal diameter.

There are tables and tables of standard size threads to choose from. Some of the most common are the unified thread standards used in the United States, and the ISO Metric screw thread used in most other countries. Both have a 60° included V shape that follows a helix around the diameter. When designing threaded components, it is more cost-effective to try and use these standard size threads. You can make up your own thread size if necessary, but it will be more costly. Gauges used to check threads of standard size are readily available and much more affordable than custom made gauges. Also, the same goes for the tools used to cut the standard size threads.

If you cannot find a standard size thread that fits your needs, then here are some things to consider (see Figure 1.15). Try and make the threads with a 60° included angle so a standard thread cutting tool can be used. Determine the major/minor diameter, pitch diameter, and thread pitch. Make sure to provide that information to the person machining the threads. A good rule of thumb is that the threads should be about 70-75% of the full thread depth. This gives sufficient thread engagement and also leaves flats on the peaks of the threads. Having flats on the peaks of the threads makes them less sharp and more resistant to damage.

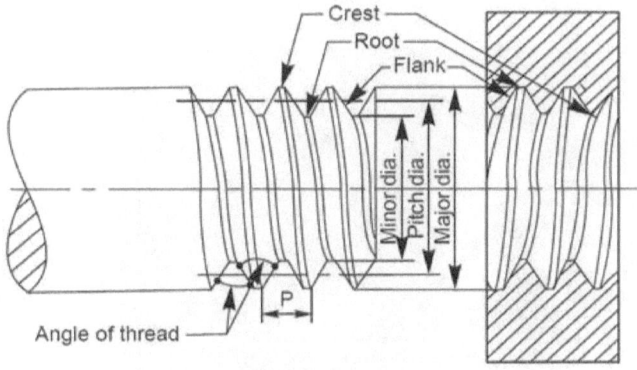

Screw thread nomenclature

Figure 1.15

43

The tolerance range of the thread including the major/minor diameter, and pitch diameters are called the class of fit. The class of fit determines how closely the parts fit together. For example, in unified threads, you have class 1, 2, and 3. Class 1 threads are a loose fit, Class 2 threads are a free fit, and class 3 are a close fit. Class 2 is the most commonly used, and they are designed to maximize strength considering typical machine shop capabilities.

Thread relief is a groove behind the thread (see Figure 1.16). On external threads, the groove diameter is smaller than the minor diameter. On internal threads, the groove diameter is larger than the thread major diameter. Manufacturers use thread relief behind the threaded diameter to provide clearance for cutting tools. It is also used to allow the thread to go all the way through the diameter. This ensures that the part can be screwed all the way down, eliminating any gaps.

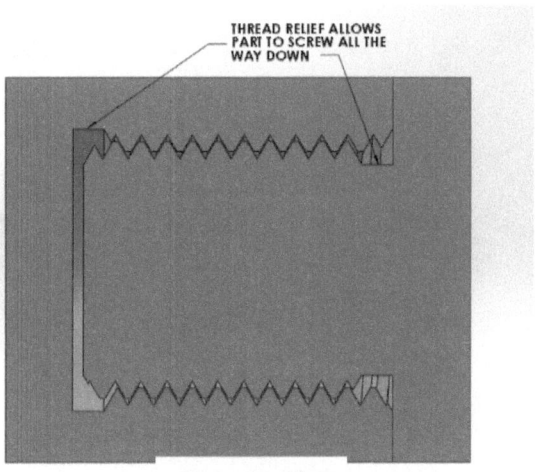

Figure 1.16

Another way of ensuring a threaded part can screw all the way down is to use a lead in diameter (see Figure 1.17). This method can be used in the circumstances where a thread relief is not possible. A lead-in diameter is a diameter before the start of the threads that is smaller than the minor diameter in the case of male threads, or larger than the major diameter in the case of female threads. This can also aid in the assembly of two threaded parts and reduce the possibility of damage to the threads. It acts as an alignment guide when you are initially starting to thread them into one another.

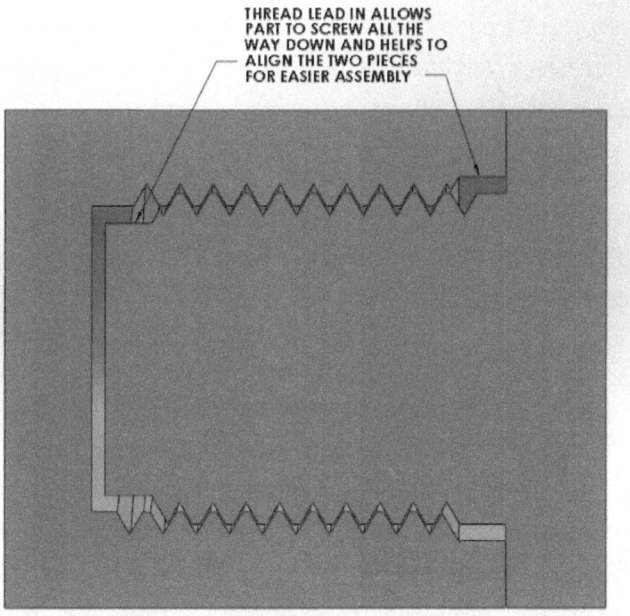

Figure 1.17

The thread pitch is the distance between two successive thread peaks (see Figure 1.15). This is controlled by the angle of the helix that wraps around the threaded diameter. Having a smaller helix angle (a smaller thread pitch) can make it harder for threaded parts to come undone. This is because the force when tightening is directed away from the direction of motion. But with a finer thread pitch, you will have to turn the bolt more times to get it tightened down. When you have a larger helix angle (a larger thread pitch) the force when tightening is directed more toward the direction of motion. This will make it easier for the threads to come undone. Although, with a coarser pitch like this you will be able to tighten the threads down faster. No matter what you decide on there is a tradeoff, it is up to you to determine what should be done for your specific application.

A reliable and complete guide for information on all types of threads is the machinery's Handbook by Franklin D. Jones and Erik Oberg. The reference book is of great help for design engineers, Machinists, and inventors.

O ring Gland Design Considerations

O-rings or gaskets help in sealing the interface between mating components to prevent intrusion or escape of fluids. O-rings or toric joints are a loop of elastomer which has a round cross-section. The O-ring gland is the groove where the O-ring will seat (see Figure 1.18). The design and shape of the gland are imperative to achieving a proper seal. Here are some design considerations that you need to pay attention to when designing the O-ring gland.

46

The cross-sectional diameter of a standard size O-rings ranges between 0.040 to 0.275 inches, whereas the inside diameter ranges from 0.029 to 25.940 inches. O-rings are called out using the inner diameter size and the size of the cross-section along with the shore scale hardness of the material. For example, for an AS568-010 size 70 shore A, Nitrile would be .239 x .070 70 BN. When designing glands for O-rings make sure that a standard size will fit properly when possible. This way you will not need to have a custom size made. The O-ring or gasket will need to sit on a smooth and unobstructed surface to ensure it seals properly.

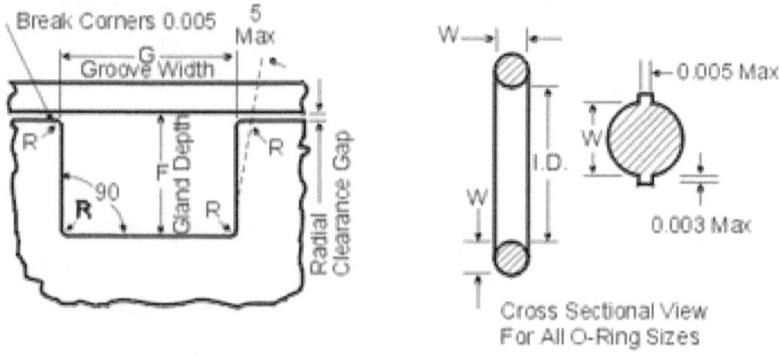

Figure 1.18

Manufacturers of O-rings such as Apple rubber products will give specifications for the glands for optimal function. You will have dynamic seals (moving seals), or you will have static seals (non-moving seals). These will give information on the amount of compression the O-ring should undergo, as well as the size of the gland so the O-ring has the proper amount of

47

room to expand under compression. Be careful to follow the advice of the manufacturer as they are experts with many decades of experience in designing O-rings and their fits.

There are several factors that you need to consider when determining the material of the O-ring. The temperature resistance, intended environment, and compatibility with chemicals are some factors that determine the material of the O-ring.

Considerations for Tolerance

Tolerances are the deviation in size from the nominal value that a part is allowed to extend to while still working for the intended purpose. The next aspect that you need to consider is the extent of the tolerance that you wish to place on your designed parts.

First and foremost, the most important part of determining tolerances is to ensure that all of the parts will fit together every time they are assembled. This can be trickier than it sounds. Of course, you want the parts to fit together, but in most cases, you want a precise fit. You do not want them too loose or too tight. This is where the importance of proper tolerancing comes into play. Some tolerancing practices can come from an experiment or experience. Also, there are standard tolerances for fits such as sliding fits, interference fits, or press fits that you can find in books such as the Machinery handbook. Pay close attention to the tolerances you place on your parts so that you are confident the fits will meet your expectations after the parts are manufactured.

This is very important. A lot of money can be wasted if parts are incorrectly toleranced and do not fit together after they have been made. This will cause costly rework or even junking the components altogether and remaking them.

Many engineers use unnecessary tight tolerances to ensure parts fit together consistently. However, the use of tight tolerances when designing a part can drive up the total cost of manufacturing. This is due to the fact that much more time and effort must be put into making them. For example, it is much more difficult to consistently machine parts that have a tolerance of ±.0005" than it is to make parts with ±.002". This is because of cutting tool wear, change in temperature, and reduction in the margin of error.

Geometric dimensioning and tolerancing or GD&T is a useful tolerancing system that can give a manufacturer more dimensional leeway but also produce a more accurate part. For example, using a GD&T positional tolerance on a hole compared to conventional coordinate tolerancing allows for 57% more tolerance while achieving the same precision (see Figure 1.19). It is a language of its own with special symbols and guidelines. GD&T allows you to not only define the size required, but also the behavior of those geometries within that specified size range. This type of tolerance is very important to know and practice in order to make consistently working assemblies and components.

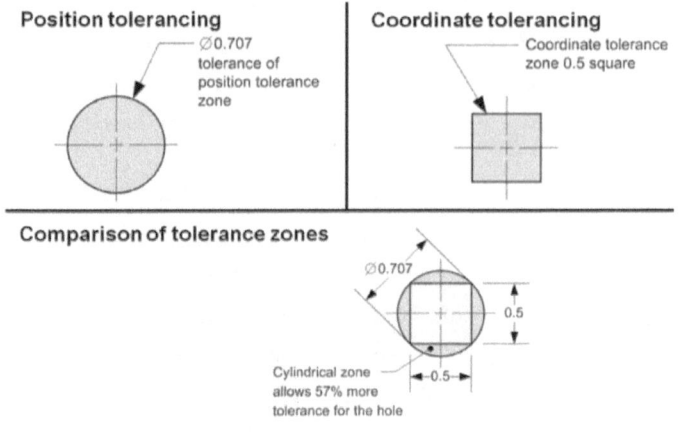

Figure 1.19

Cost Considerations

Another important consideration that you need to focus on before you begin a design process is the cost involved. It is important to consider the cost of the selected materials, the manufacturing process that will be used, and the interchangeability of components.

The size of the parts you use affects the overall cost. A smaller part, for example, requires less material and is considerably cheaper. It is always a good practice to make parts with as little material as possible to reduce material costs.

One good way to reduce the costs of the whole design is to consider the interchangeability of the parts you are designing. Try and make parts that can be used multiple times for different functions. Consider the fact that you may be able to use a certain part in more than one assembly. Be creative with parts that already exist, and try and incorporate them into your current design rather than having to manufacture an entirely new one. This will help to keep the number of different components you have in inventory down. Also, it will reduce the time and material costs associated with making another part.

It is almost always cheaper to use off the shelf parts than to invent and make something new. Before you go inventing something new, look around and see if there is already something out there that will fit your needs. There are times when you can find parts that someone has in stock that are just right for your assembly. More than likely the places have mass-produced these parts and will be able to sell it for much less money than it would cost you to make them.

Look for ways to make parts that are easy to mold, manufacture, and machine. Simple is better, so simplify your designs and assemblies as much as possible. You can often achieve the same end results with less.

Consider the Strength
Manufacturers often define the strength of the part and the role and nature of work it is required for. Based on the intended use of the designed part, you should ensure it has the proper strength and durability to last.

A modern way of checking the strength of parts is Finite Element Analysis (FEA). This is done through designing software such as SolidWorks, Creo, Fusion 360, and many others. FEA sections parts into tiny finite pieces and individually tests the strength of those pieces using. It takes into consideration the arrangement of the geometry and the relationships between each tiny finite piece under a load. It will give you results that can help you in optimizing the components during the design phase.

There are various material properties that you need to be aware of for carrying out an accurate fine element analysis. The bulk and shear modulus, the Poisson ratio, yield strength, tensile strength, and the density of the material are concepts that can help you in carrying out an FEA.

Be aware of sharp corners where high levels of stress and strain are present. Sharp corners are stress concentration points. Thus, it is wise to add as large a radius as possible in these areas.

Considerations for assembly

One key aspect of the design process is assembly and like all the other stages, there are certain factors that you need to be mindful of when designing components that will need to be assembled.

Firstly, you need to ensure that it is easy for you to assemble and disassemble. Part of this goes back to proper tolerancing of your parts to ensure they assembly with little effort.

Unless of course if you want press-fitting or some other similar form. You must put yourself in the shoes of the person who will be doing the assembly. Mentally walk through the steps that he or she will take to assemble your parts. Design your assemblies in such a way that even the least knowledgeable person in the assembly department can do it fairly quickly. Ease of assembly can save a lot of time and money when it comes to production. Little details that you add or remove to make assembly easy can add up to hours, days, or even weeks of saved time.

Avoid designing parts that would require special or custom tooling to assembly. For example, fasteners with common drives such as hex head or Philips head screws can save time and money. Those tools are readily available and easy to locate rather than making or searching for a special tool.

Consider how the parts will be held in order to assemble. If the parts must be screwed together by hand and there is nothing to grip on consider adding a knurl or wrench flats. A knurl is a pattern of diamond-shaped grooves that span the circumference of a cylindrical part. This makes it much easier to grip and turn with just your hands.

In the scenario that you have an O-ring or gasket, make sure there is a way to easily compress it for a proper seal. There should be an even amount of force applied to the entire length of the gasket. This helps reduce high or low-pressure spots and helps it seal completely. In the case of O-rings, it is recommended that they are installed with lubrication to avoid fatigue and help with installation. Do not obstruct the sealing surface with holes or any other feature that could cause the seal to be lacerated or fail. Furthermore, make sure that an O-ring has a clear and non-compromising path to its point of installation. For example, you wouldn't want to stretch and roll an O-ring over sharp corners or threads during assembly. It could possibly be torn or cut and you may not even know it until the seal fails. By that time, it is too late and it may be a serious problem.

Plan on all the sharp edges of the components being chamfered. A chamfer is a small angle, usually 45° on edges that eliminates any burr. It also makes them less sharp so that nobody cuts themselves while handling or assembling.

When making considerations for assembly, you need to consider that a low-paid employee of the manufacturer may be responsible for the final assembly. Therefore, it is your job to ensure the assembly stage is as simple as possible and doesn't require any extraordinary skill.

Assembling Model Components in Software

This section will focus on assembling the components inside of your design software. 3D modeling software such as SolidWorks, Fusion 360, and Creo all have different design environments for creating individual parts and then making 3d assemblies from those parts. Usually, you will first design the parts by themselves. Then you can mate them together in the assembly environment, or even build other components around existing ones to get your assembly.

There are certain procedures that one can follow that will significantly reduce time and allow for more flexibility when analyzing the assembly.

First, it is important to build your assembly starting with the most logical base part. A lot of times this would be the part that you would start with when assembling in real life. This will be the foundational part of the design where others can be added and built up around it. For example, if you were assembling an engine, the most logical piece to start with would be the engine block. You would then start adding components around it such as the crankshaft, bearings, pistons, etc.

It is always a good idea to use the x,y,z origin in the model space to begin assembling around (see Figure 1.20). This will help you take advantage of the symmetrical nature of most designs. Doing it this way you can easily cut the assembly down the middle to get a clear view of the entire cross-section of the assembly. This will help you when it comes time to inspect your design for flaws, and for fit and function. Also, this gives more flexibility if you need to use patterning features to add multiple copies of the same components such as a circular screw pattern.

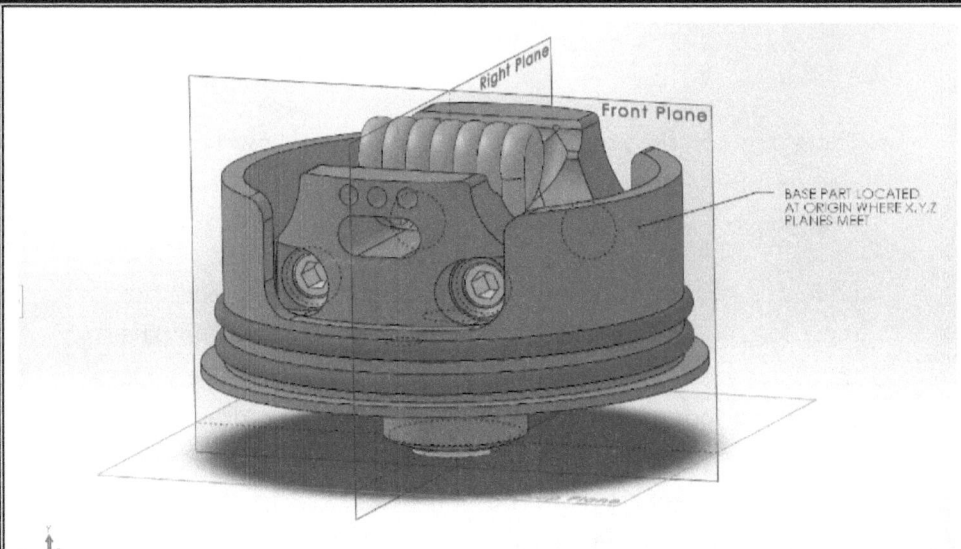

Figure 1.20

The transparency feature is an awesome tool to give you a viewport into the internal workings of your assembly (see Figure 1.21). This tool allows you to make parts transparent so you can see inside the assembly to evaluate the relationships of the internal parts. It also helps you to get a sense of what the components will look like when they are held captive internally where you would never normally see them in real life. This is a very useful tool for ensuring your internal cavities provide enough space for items such as wiring, connectors, or screw heads.

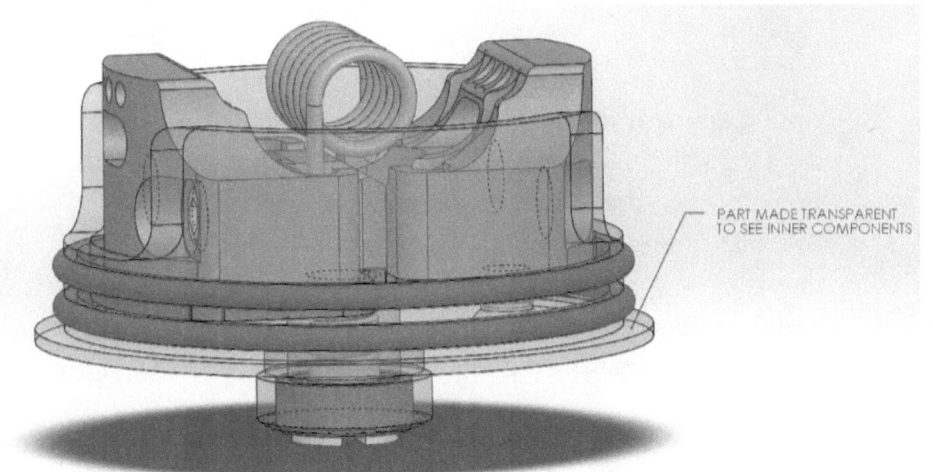

PART MADE TRANSPARENT
TO SEE INNER COMPONENTS

Figure 1.21

It is sometimes very convenient to make edits to parts directly in the 3d assembly. This gives you the advantage of being able to see how all the parts interact while changing the geometry of individual components. This can also be a time saver since you are not flipping back and forth between parts and assembly. When you edit or change the geometry of one part in an assembly you don't need to rely on memory because you can immediately see how the change affects the entire design.

When I am designing new parts around existing parts I like to have 3d models of everything. For example, I was tasked with designing custom jaws that would hold different size tubes to be cut at different angles. These jaws needed to mount to the vise on our large metal cutting chop saw. The first thing I did was carefully measure the vise and draw a 3d model of it. I then designed the jaws and assembled them in SolidWorks. That way I could see it in its entirety, evaluate its fit, form, and function and determine it would work properly prior to cutting any real metal. To some, this may sound like extra work that isn't necessary. What comes to my mind is the old saying of measure twice and cut once. It may be more work and take more time, but it gives you a better chance at thwarting any problems ahead of time. You will also have more confidence that everything works seamlessly when it comes time for making the real components.

The reason you design in 3d is so you can visualize a product before you make it. They say a picture is worth a thousand words. So be sure that all the components are the correct colors and have the correct material properties applied to them. Having the correct colors will give you a more accurate visual representation of the real-life version. Having the correct material properties will give you a reference when it comes time to build. When you apply material properties as you go, it also can save time later if you decide to do FEA on the parts as it is a requirement.

File Management and Organization

Like any other professional work on software, it is important to manage your files effectively. Over many years and hundreds of designs your files can easily become large and unorganized. It is of utmost importance that you take great care choosing the names and locations for files on your computer. You may have something you worked on years ago that needs to be called on. It can save you tons of time if you know exactly where it is and can retrieve it quickly.

One good way to name your files in an organized manner is by using similar prefixes for the names of the parts you design. Let's imagine you have an assembly and you have named it the REAR HUB. Every part in that assembly should start off with the words REAR HUB and then it's specific name after. Create specific folders for every assembly you create. Once again, name the folder REAR HUB so you can immediately identify it.

If you are not using any sort of product data management software (PDM) it can be helpful to add the revision number into the title of the part. This way, any new revisions can be saved with new numbers and you can always go back to old versions if you need to. Also, make sure the revision numbers for the models match the numbers for the blueprints (see Figure 1.22). For example, the first design of a part could be called "REAR HUB – BRACKET REVA-01".The title of the corresponding blueprint would be the exact same name. The revision number in the revision table, as well as in the title block, would be Revision A-01. If you make a change then save it as a new part named "REAR HUB – BRACKET REV A-02" and so on and so forth. If you are using a product data management such as SolidWorks PDM then this is irrelevant. The PDM will automatically change revision numbers, keep past revisions, and store all the associated data.

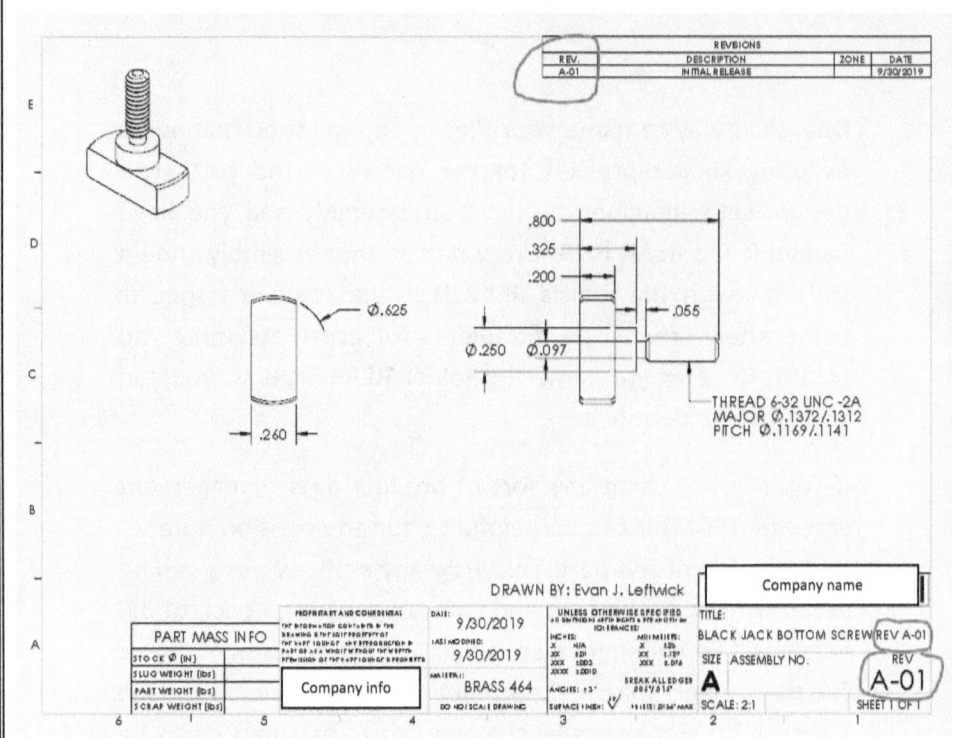

Figure 1.22

Outsourcing/Finding Good Information for Design

There will undoubtedly be times when you need to look things up or find further information on a certain subject when creating new designs. It is impossible to know everything. Nowadays we have the most powerful information tool the world has ever known, the internet. With all the great information available on the internet, there is also a lot of bad information on there too. When designing products, you must make sure that the sources you are choosing to get your information from are reputable and trustworthy.

60

Books such as the Machinery Handbook have been around for decades and are a very trustworthy source. There are many others but make sure to do your own research on the authors and determine that they are reliable sources of information before you stake your career on it.

The digital age and easy access to information have further simplified the process. Trustworthy internet sources and credible researches uploaded on the internet are other options that you can consider for getting the right answers. Websites like engineering edge are a credible source of information and are of great help for the user.

The American Society of Mechanical Engineering is another platform where you can get many answers to your questions. Becoming a member of the platform allows you to connect with thousands of other engineers from the country. You can also get certification and accreditation using the website.

Manufacturing companies that specialize in certain areas usually have a lot of good information directly on their websites. These are always good sources of information since it is their area of expertise and more often than not, they possess decades of experience to back it up.

Creating Blueprints for Manufacturing

Creating Blueprints for manufacturing is one of the most tedious and important parts of the design process. It should be viewed as a binding contract between you and the manufacturer of the parts. It is your job to include every detail, dimension, and feature that you expect to be on the finished product. If you fail to specify something that you want on the blueprint, they have absolutely no obligation to provide it to you. This is why you must pay extra close attention.

When making a blueprint ensure that it is done on a professional-looking drawing sheet format (see Figure 1.22). This means that it should have a border, title block with proper information, and revision table. The title blocks tend to vary from company to company but typically they include the drawing sheet size, part title, designer name, company name, date, revision number, drawing scale, and unless otherwise specified (UOS) tolerances. Also, ensure that the revision number in the title block matches the revision number in the revision table.

The orderliness of the dimensioning scheme is another crucial part of creating a blueprint. You need to ensure that the dimensions look orderly and clean, enabling the person reading it to easily interpret the required dimensions.

Blueprints are a beautiful integration of three important elements. The drawing, notes, and dimensions. The orderliness of the dimensioning scheme provides you with information about the tolerance levels for each feature.

Utilize drawing techniques such as cutaways, crops, section views, and detail views to help you accurately convey all geometry and dimensions of a part. These illustrations help you in avoiding confusion related to the project, providing them with the much-needed contrast between certain features.

Chapter 4- In-House Manufacturing Considerations

In-house manufacturing considerations apply when you work for a company that will be making its own parts in the house. This employs the intellect of various stakeholders in your company and the departments they are associated with.

Firstly, you need to ensure that an accredited coworker double-checks the design for flaws and errors. Scrutiny and critical reviews are aspects of the design process that can elevate the standard of your design. Be willing to take advice from the people responsible for manufacturing the part.

Other people in your company can help you make design changes that are revolutionary for the product. You should also consider advice from people who are responsible for assembling and making parts. Since they do it on a daily basis, they may have valuable insight into ways to make the components easier to assemble. Communication with the assembly and manufacturing team can help you identify tiny details that can hinder the manufacturing process.

The people responsible for assembly should also be aware of your design intent. Clear communication of the desired end product should be explained to them. Provide them with accurate and concise assembly drawings so there is no confusion on how to put the parts together. Remember that they do not have the same intimate knowledge of the product as you do. They are relying on you to provide them with the necessary information to do their job.

Chapter 5- Sending Blueprints and Solid Models

On many occasions, you will be sending the design out to be made elsewhere. In these modern times, almost manufacturers use some sort of 3D modeling software to aid their processes. This will allow them to see it in 3D just as you do. This can be greatly helpful in conveying your design intent. You will need to provide clear and accurate instructions on what you expect, just as when you are manufacturing in house. Here are some factors that you need to consider.

When sending the components to the manufacturer it is invaluable to include both the 3D model and the Blueprint. The 3D model will allow them to visualize the part and easily see features that may be hard to decipher in just a 2D drawing. Many manufacturers can make machine programs for the parts directly from the 3D model. Once the part is in the process of being made, shop personnel will use the blueprint to ensure the part meets the customers' standards. This is why it is important that the dimensions of the 2D blueprint must match the dimensions of the 3D model to the T.

The dimensions in the 3D model, as well as the blueprint, should be on the nominal dimensional value. For example, let's assume that you have a 1.000" diameter with a limit tolerance of 1.005"/.995" on the blueprint. You want the dimensions of your 3D model to be exactly 1.000" since this is the nominal value. This may sound obvious, but there may come a time when you have a dimension of 1.002" and you decide to put a bilateral tolerance or limit tolerance that still would put the part in the range of 1.005"/.995". Try and avoid doing this. Remember, many times the manufacturer will be making a machine program directly from the model. If you do not have your model dimensions on the nominal, the program they make will not be on the nominal. The results could mean extra work for the shop personnel and adjustments to the program.

As stated in the previous section, ensure that your blueprints and models exhibit proper revision control. Companies will keep saved blueprints for work they have done prior. If you make a change to a part and don't change the revision of the blueprint, they will not know it is different. They may just pull up the old blueprint and use that to make the part. This would result in inaccurate parts and the fault would lie squarely on your shoulders.

Manufacturers can be very lenient and willing to work with you to ensure you get what you need. Firstly, they want you to be a happy and satisfied customer, and second, they do not want to have to make the same parts twice. So be sure to communicate with them on what you are trying to achieve. If you have other parts that need to fit together with the ones they are making, you should bring that up to them. I have worked with companies that have allowed me to give them samples so they can ensure a proper fit while they are making them. This will help alleviate any errors on your side or theirs.

Ensure that the tolerances you are trying to achieve are within the limits of the manufacturer. For example, let's say you are trying to achieve a ±.0001" tolerance on a machined part. Speak to them first and make sure that they are able to achieve that with the equipment they have. Most shops won't turn down work, what they will do is charge you more money to compensate for the level of difficulty.

Chapter 6- Receiving Parts

The last leg of the whole design process is to check the final manufactured parts. As an authority of the design process, your job is to confirm whether they were made according to your specifications.

Upon receiving parts that were made by and outside vendor it is important to check them for accuracy. Take a random sample and carefully measure or inspect them to ensure they meet the quality you desired. If they were not made to the specifications you instructed, it is much easier to act and fix immediately, rather than finding the problem later down the road.

Given that you were responsible for designing the components, it won't be tough for you to spot undesired changes. This will make it easier for you to review the final parts and come up with a solution to the problems. Remember the design intent and thoroughly validate they are working the way you planned before accepting.

Conclusion

Design Engineering can be an extremely challenging yet rewarding career. It is possible to systematically identify your design goals, then strategically find solutions to attain your desired results. The feeling of accomplishment after attaining these positive results is likely to last a lifetime.

I strive to offer other designers, inventors, and engineers out there with help in doing their jobs successfully. My hopes are that with this guide they will be better equipped to tackle the real-world challenges that lie ahead. In addition to giving a step by step guideline of the design process, this book provides you an insight into many of the important details to be aware of in order to be a prosperous designer.

O